To the Frahm Pike Houtses,
past and present.
– *M.H.*

For Mom and Dad,
for everything.
– *J.B.*

Text © 2021 Michelle Houts
Illustrations © 2021 Jen Betton

ISBN 978-1-948898-05-8

Library of Congress Control Number: 2021933955

Designed by Mary A. Burns   Edited by Emma D. Dryden

Printed in the United States of America
First Edition
10 9 8 7 6 5 4 3 2 1

Barn at Night

By Michelle Houts
Illustrations by Jen Betton

FEEDING MINDS PRESS

American Farm Bureau Foundation for Agriculture®

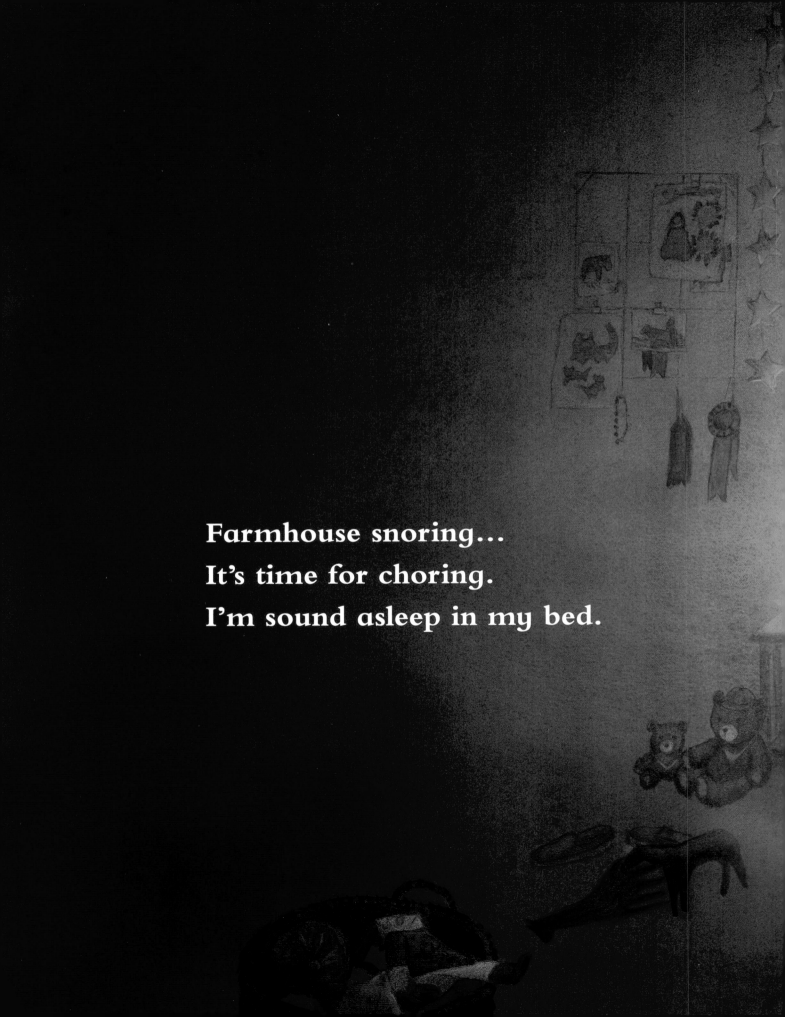

Farmhouse snoring…
It's time for choring.
I'm sound asleep in my bed.

Gently waking;
Shivering, shaking.
"Let's get to work, sleepy head!"

Up before dawn.
A stretch and a yawn.
Morning comes fast on the farm.

The sky's dark as night
in setting moonlight.
We tiptoe out to the barn.

Crisp winter air.
Chilly wind in my hair.
Across the barnyard we go.

The only sound
for miles around –
our boots on the
crunchy white snow.

With a heave and a shove,
the door high above
slides noisily off to one side.

Turning to greet us
are Ed and Miletus
as we slip quickly inside.

Glowing green eyes—
Oh! What a surprise!—
disappear as we turn on the light.

We've frightened a guest,
an unwelcome pest
who snuck in the barn late last night.

A curious steer
pricks up one ear.
Mule lifts his head back and brays.

We've let in the cold
we're now being told
in the kindest of animal ways.

Tubby cats scurry.
What's the big hurry
to sit by the milkpan and wait?

Hopeful round eyes,
hungry cat cries,
as calves stomp their hooves by the gate.

Straw for a bed—
hay stored overhead—
molasses and milk smell so sweet.

Cracked corn and grain—
soft fur and mane—
the smells in the barn are unique.

Waiting in line,
one at a time,
the calves slurp a milky warm brew.

Mouths full of hay,
the big cattle play,
then stand still and happily chew.

In a dark corner stall,
against the back wall,
stands Eleanor, patient and calm.

A nicker, a nuzzle,
her velvety muzzle
nudges and tickles my palm.

When darkness falls
and animals call,
we'll go out and feed them again.

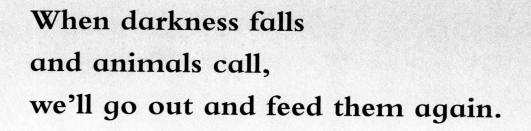

Chores in the light
are adventures at night
for me and my animal friends.

Late one winter night,
I notice a light.
I silently slip down the stairs.

"Oh, can I come too?
I know just what to do!"
We step into the frosty night air.

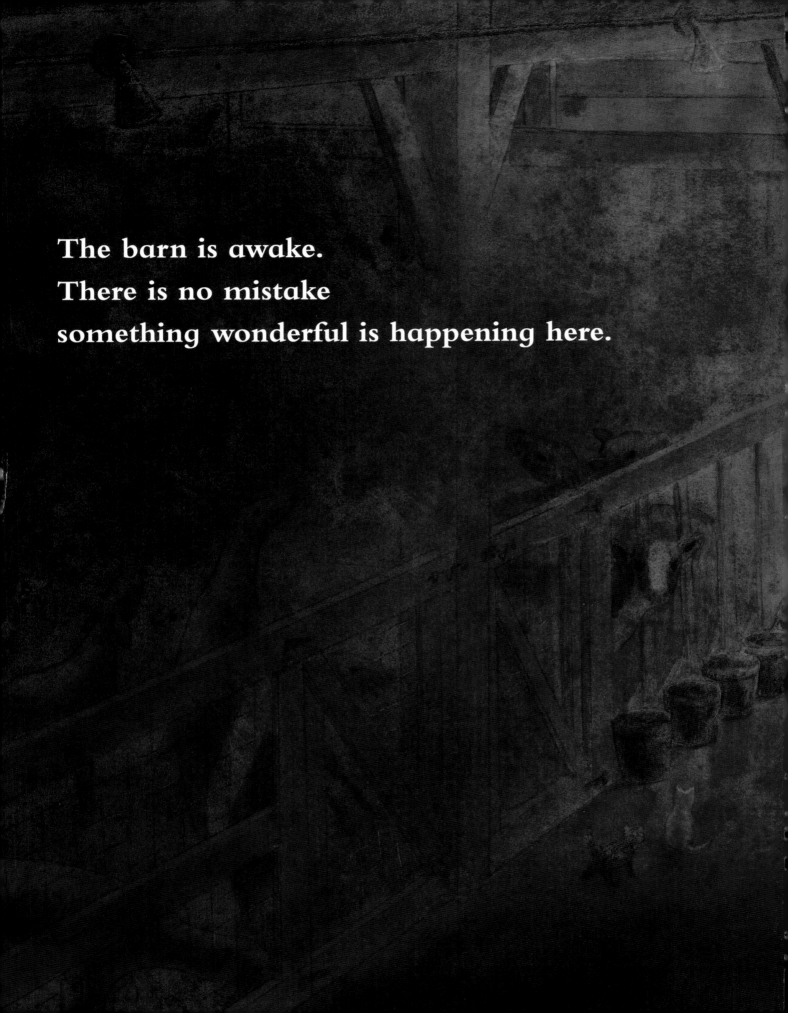

The barn is awake.
There is no mistake
something wonderful is happening here.

Every eye shows it.
Each animal knows it,
every calf, every goat, every steer.

A whinny so loud.
A momma so proud.
Eleanor shows us her foal.

We enter the stall
and there on the straw
he lies shivering, wet, black as coal.

I don't move an inch,
not daring to flinch.
Holding my breath as we wait...

On wobbly knees,
with trembling ease,
he takes his first step toward the gate.

Yellow panes glowing,
it begins snowing.
Over rafters a hoot owl takes flight.

A safe place to dwell—
all here is well—
when we're in the barn at night.